Commercial and Military Applications and Timelines for Quantum Technology

EDWARD PARKER

Prepared for the Department of Defense
Approved for public release; distribution unlimited

NATIONAL DEFENSE RESEARCH INSTITUTE

For more information on this publication, visit **www.rand.org/t/RRA1482-4**.

About RAND

The RAND Corporation is a research organization that develops solutions to public policy challenges to help make communities throughout the world safer and more secure, healthier and more prosperous. RAND is nonprofit, nonpartisan, and committed to the public interest. To learn more about RAND, visit www.rand.org.

Research Integrity

Our mission to help improve policy and decisionmaking through research and analysis is enabled through our core values of quality and objectivity and our unwavering commitment to the highest level of integrity and ethical behavior. To help ensure our research and analysis are rigorous, objective, and nonpartisan, we subject our research publications to a robust and exacting quality-assurance process; avoid both the appearance and reality of financial and other conflicts of interest through staff training, project screening, and a policy of mandatory disclosure; and pursue transparency in our research engagements through our commitment to the open publication of our research findings and recommendations, disclosure of the source of funding of published research, and policies to ensure intellectual independence. For more information, visit www.rand.org/about/principles.

RAND's publications do not necessarily reflect the opinions of its research clients and sponsors.

Published by the RAND Corporation, Santa Monica, Calif.
© 2021 RAND Corporation
RAND® is a registered trademark.

Library of Congress Cataloging-in-Publication Data is available for this publication.

ISBN: 978-1-9774-0752-8

Cover image: Graham Carlow/www.grahamcarlow.com.

Limited Print and Electronic Distribution Rights

About This Report

This report is one of a set of focused, short analyses that examine critical emerging technologies and provide perspective on their development around the world. The analyses cover quantum technology, technologies for enhancing human performance, semiconductor technology, the intersection of artificial intelligence and cybersecurity, deepfake generation and detection, and the use of patent data to assist in understanding global trends in emerging technologies. This report focuses on quantum technology.

The short technology analyses draw from open sources to provide a snapshot in time of the status of the technology; its plausible evolution; and the thought leaders, firms, institutes, and countries that are working on it (with particular interest in the prospects of potential adversaries of the United States). These examinations of individual technologies also provide background information useful for examining patents and scientific articles for insights in how a technology may evolve, its implications, and who will lead in its evolution. They were developed quickly to respond to the sponsor's requests for rapid completion of focused, open-source literature reviews; by design, they do not address possible policy implications.

The analysis of patent data is different from the other assessments in that it both demonstrates an analytical approach and applies it to two specific arenas: military technologies and semiconductors (included as an appendix to the brief semiconductor analysis). That analysis illustrates how patent data can provide insights into the evolution of technologies via the data associated with the patent's inventors, their organizations, and their country. Together, the authors of the reports in this series attempt to foresee the implications of emerging technologies via different data sources in a systematic and transparent fashion.

The research reported here was completed in July 2020 and underwent security review with the sponsor and the Defense Office of Prepublication and Security Review before public release.

RAND National Security Research Division

This research was sponsored by the U.S. Department of Defense and conducted within the Cyber and Intelligence Policy Center of the RAND National Security Research Division (NSRD), which operates the National Defense Research Institute (NDRI), a federally funded research and development center sponsored by the Office of the Secretary of Defense, the Joint Staff, the Unified Combatant Commands, the Navy, the Marine Corps, the defense agencies, and the defense intelligence enterprise.

For more information on the RAND Cyber and Intelligence Policy Center, see www.rand.org/nsrd/intel or contact the director (contact information is provided on the webpage).

Acknowledgments

The author would like to thank the principal investigators for this project, John Parachini and Marjory S. Blumenthal, and the quality assurance reviewers, Chad J. R. Ohlandt and Charles Brown.

Summary

The physical theory of quantum mechanics has been studied for more than a century, but in recent years, many new potential applications have been proposed and have entered or approached practical development. Although some of these applications are much closer to readiness than others are, many of them could be important in both the commercial and the national security spheres.

Quantum technology is grouped into three broad categories: quantum sensing, quantum communication, and quantum computing.

Quantum sensing has a broad variety of potential applications, such as biomedical imaging, enhanced imaging and radar, and navigation in environments where use of the Global Positioning System is denied. Some of these applications could be ready within a few years.

The primary near-term application of quantum communication is to improve the security of communications against interception and eavesdropping (sometimes referred to as "unhackable" communication in the popular press, although there are still security vulnerabilities in practice). This technology has already been commercially deployed to a limited extent in Europe and to a greater extent in China, but not in the United States. Many U.S. experts are skeptical that quantum communication will provide a significant improvement in secure communications. Quantum communication also has applications beyond enhancing security (such as networking together quantum computers), but these other applications will not be ready for many years.

Quantum computing is significantly further away from being technologically ready than quantum sensing and quantum communication are, but it would have the greatest long-term impact if successfully deployed. Quantum computing could eventually lead to major improvements in biochemistry, materials science, drug discovery, and (further into the future) machine learning. More concerningly, a quantum computer that is greatly scaled up from current sizes could eventually threaten the security of the public-key encryption that is currently used to protect virtually all data sent over the internet. However, major technological progress needs to be made before any existing quantum computers reach these theoretical capabilities, and it is not clear what applications (if any) near-term quantum computers will have.

Most industrialized countries invest significant resources for research and development into quantum technology. The United States and China lead in overall investments, but Canada, the United Kingdom, the European Union, Japan, and Australia are all significant players. Broadly speaking, the United States is the world leader in quantum computing and possibly sensing, while China is the world leader in quantum communication. For example, private-sector U.S. companies have built proof-of-principle quantum computers that are at or near the point of surpassing the world's most-powerful supercomputers at certain extremely specialized calculations, while China is the only country that has successfully launched a satellite capable of transmitting quantum-secure intercontinental communication, albeit at limited amounts.

Contents

Figures and Tables

Figures

Tables

CHAPTER ONE

Introduction

Quantum mechanics is the field of physics that describes the behavior of microscopic particles. At distance scales smaller than a few nanometers (10^{-9} meters), fundamentally new physical effects become important that have no analog with humans' everyday experience. Engineers are now beginning to be able to create several types of devices that take advantage of these effects in order to far exceed the capabilities of existing devices.

Quantum technology is a broad umbrella term that covers these types of devices, many of which are still in the early experimental stage. Quantum technologies are often grouped into three broad categories: quantum sensing, quantum communication, and quantum computing. Quantum sensing is the closest to useful commercial deployment but the least potentially disruptive, while quantum computing is the furthest from commercial deployment but the most potentially disruptive. Palmer, 2017, gives a nontechnical overview of the potential near-term uses of all three types of quantum technology as of 2017, although there have been significant developments since then.

This report briefly describes the current state and prospects of quantum technology in each of these areas.[1] It also considers the likely commercial outlook over the next few years, the major international players, and the potential security implications of these emerging technologies.

Table 1.1 gives a summary of the areas of quantum technology, their applications, predicted timelines, and the current national technology leaders.

---

[1] This research was completed in July 2020.

TABLE 1.1

Summary of Quantum Technology Areas

Quantum Technology Area	Quantum Sensing	Quantum Communication	Quantum Computing
Description	• Enhanced-sensitivity sensors for time, acceleration, magnetic fields, and electromagnetic radiation	• Transmission of signals that use the quantum properties of light	• New computing technology that can simultaneously process large amounts of information
Example applications	• Inertial navigation systems and magnetometry systems for navigating in GPS-denied environments • Improved lidar and radar for ISR	• Secure communications that are difficult to intercept • Networking together of quantum sensors and computers	• Design of advanced materials • Biochemistry • Drug design • Numerical optimization (e.g., for logistics) • Encryption-breaking
Predicted time frame	• Potentially the next few years for navigation • 10+ years for improved radar	• Communications security applications already commercially available in China, the EU, and Japan • Networking applications still many years away	• Possible niche applications within 5 years • Most important applications likely 10 years away
Current world leaders	• Little public reporting, but the United States and the United Kingdom are both strong	• China, followed by Japan, South Korea, and the EU	• United States, followed by Canada and the EU

NOTE: EU = European Union; GPS = Global Positioning System; ISR = intelligence, surveillance, and reconnaissance.

CHAPTER TWO

Overview of Quantum Technology

This chapter gives an overview of potential uses for each of the three main categories of quantum technology: quantum sensing, quantum communication, and quantum computing.

Quantum Sensing

Quantum sensing and metrology (Degen, Reinhard, and Cappellaro, 2017; Bowler, 2019) refers to the ability to use quantum mechanics to build extremely precise sensors. This is the application of quantum technology considered to have the nearest-term commercial potential (Palmer, 2017).

Precision Measurement of Time; Acceleration; and Electric, Magnetic, and Gravitational Fields

One category of sensing technology includes probes that can perform highly sensitive measurements of elapsed time; acceleration; or electric, magnetic, or gravitational fields. Highly precise clocks could have commercial applications, such as verifying high-frequency financial trades and dynamically regulating a smart electrical grid. Sensitive gravitometers could be used for seismographic predictions of earthquakes and volcano eruptions, for underground exploration of oil and gas reserves, or for evaluations of the solidity of the ground beneath major construction projects without needing to dig boreholes (Bowler, 2019).[1] The German company Bosch, which man-

---

[1] The environmental consulting firm RSK has estimated that up to one-third of major construction projects experience major delays because of unexpected underground conditions (Palmer, 2017).

ufactures automobile components and other precision industrial technologies, is studying the possibility of incorporating quantum accelerometers in automobiles, particularly autonomous vehicles. Quantum magnetometers could also have biomedical applications—for example, for improving magnetic resonance imaging (MRI) and positron emission tomography (PET) scanners (Palmer, 2017).

Another application of quantum sensing, primarily of interest to the military, is for navigation in GPS-denied environments (Kramer, 2014; Tucker, 2014). The Earth's gravitational and magnetic fields have small variations known as *anomalies* that vary from place to place. A sensitive gravitometer or magnetometer could take precise measurements of the local fields and compare them with maps of these anomalies to navigate without need for any external communication (Jones, 2014). Similarly, improved accelerometers could allow for self-contained inertial navigation systems (INSs). One defense laboratory researcher in the United Kingdom estimated that these positioning, navigation, and timing (PNT) applications of quantum technology could allow a submerged submarine to fix its position 1,000 times more accurately than is currently possible (Marks, 2014). Because this navigation technique is based entirely on fields inside the operating vehicle, it requires no external communication, cannot be intercepted, and is extremely difficult to jam. However, one disadvantage of magnetic-field navigation is that the local magnetic fields need to be mapped out ahead of time in the region to be navigated, which could be challenging in denied areas.

Quantum Imaging

Another category of quantum sensing is known as *quantum imaging* (Genovese, 2016; Pirandola et al., 2018), which includes the detection of both visible light and radio-frequency radiation (i.e., radar). Two of the most-promising applications are known as *ghost imaging* and *quantum illumination*.

Ghost Imaging

Ghost imaging uses the unique quantum properties of light to detect distant objects using very weak illumination beams that are difficult for the imaged target to detect (Meyers, Deacon, and Shih, 2008; Shapiro, 2008; Hardy and Shapiro, 2013). Ghost imaging beams can also strongly penetrate through atmospheric obscurants, such as smoke and clouds (Meyers et al., 2012). Researchers at the U.S. Army Research Lab, who developed the technology, have proposed using ghost imaging for long-distance ISR from the ground, unmanned aerial vehicles, or satellites. In addition to its clear military applications, the capability to penetrate smoke and cloud cover could be useful for weather monitoring or disaster response (for example, monitoring the spread of wildfires). This technique could be particularly effective through smoke or strong atmospheric turbulence and has been publicly demonstrated over a distance of 2.33 km (Army Research Laboratory, 2013). Figure 2.1 depicts the results of an earlier demonstration over a distance of 1 km.

Morris et al., 2015, and Aspden et al., 2015, discuss the possibility of using ghost imaging for covert operations. Ghost imaging can offer higher resolution than standard imaging can, particularly in very low-light environments, so the illumination beam can still produce a visible image even if the beam is much weaker than would be required for standard imaging; therefore, the beam is difficult for the target to detect.[2] Moreover, even if the signal beam is detected, the encoded image cannot be intercepted, because some of the information required to reconstruct the image is stored within the ghost imaging apparatus (Meyers, Deacon, and Shih, 2008). Morris et al., 2015, and Aspden et al., 2015, also propose that very low-illumination ghost imaging could be used to image biological samples that are so delicate that they would be damaged by bright illumination.

Quantum Illumination

Quantum illumination (first proposed in Lloyd, 2008) is conceptually similar to ghost imaging but could provide an even greater sensitivity improvement. Tan et al., 2008, mathematically proved that a quantum illumination apparatus could attain a signal-to-noise ratio (SNR) that is 6 dB higher

---

[2] However, because of the very low light levels, the tabletop demonstrations described in these articles required very long exposure times of 30 seconds to 30 minutes for high image quality, which might limit the utility of this technique for imaging moving targets (Aspden et al., 2015). However, practical applications might use higher illumination levels, which would decrease the required exposure time at the cost of an increase in detectability. Low-illumination ghost imaging has been demonstrated at both visible-light and infrared wavelengths.

FIGURE 2.1

Ghost Imaging of Two Automobile License Plates Located in a Window 1 km Away from the Camera

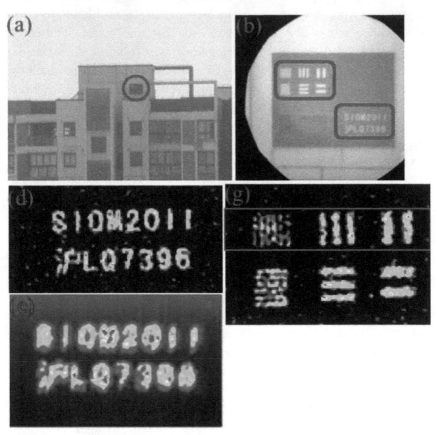

SOURCE: Adapted from Chengqiang Zhao, Wenlin Gong, Mingliang Chen, Enrong Li, Hui Wang, Wendong Xu, and Shensheng Han, "Ghost Imaging Lidar via Sparsity Constraints," *Applied Physics Letters*, Vol. 101, No. 14, 2012, Figure 2, p. 141123-2, used with the permission of AIP Publishing.

NOTES: Panels (d) and (g) show the images from the ghost imaging apparatus; panel (e) shows the image taken by a conventional camera. Some images in the original figure have been removed to save space. The character at the bottom-left corner of panels (d) and (e) is a Chinese character.

than the best *theoretically possible* apparatus that does not take advantage of quantum mechanics.[3] One proposed use case for this technology is a *quantum radar* for military applications (Ball, 2015; Barzanjeh et al., 2015). In theory, quantum radar should be uniquely effective for detecting low-reflectivity targets against a high-noise background, so it could be particularly effective for detecting stealth aircraft (Seffers, 2015). A 6-dB improvement in SNR corresponds to a 41-percent increase in the radar's maximum detection range.[4] Moreover, quantum radar could, in principle, be designed so that the signals would be extremely difficult to intercept or jam (Shapiro, 2009).[5]

No large-scale quantum illumination radar has yet been reported in the open literature, but there have been demonstrations of tabletop prototypes (Luong et al., 2019). One such prototype successfully demonstrated an SNR higher than possible for a non-quantum device (Zhang et al., 2015).[6]

As with ghost imaging, the enhanced sensitivity of quantum illumination could allow for the use of a much weaker signal beam than is typically possible. Quantum illumination could therefore also have applications for biomedical imaging in situations in which the strong signal beam from a conventional imaging device would damage the tissue being studied (Ball, 2015; Seffers, 2015).

Even the theoretical design of quantum radar remains at an extremely early stage, and until recently, quantum radar was capable (in principle) of improving the measurement of only a target's range from the transceiver but not the target's direction. However, a 2020 scientific paper improved the theoretical design to be able to simultaneously improve the determination

---

[3] However, one important caveat is that the best currently known receiver design can attain only 3 dB of this theoretical 6 dB improvement (Guha and Erkmen, 2009).

[4] A 6 dB improvement in SNR corresponds to a multiplication of SNR by a factor of 4. Because the radar range equation gives that SNR scales as the fourth power of the maximum radar range, a fourfold improvement in SNR changes the maximum radar range by a factor of $4^{1/4} \approx 1.41$, a 41-percent improvement.

[5] As with quantum radar, some of the information required to analyze the reflected signal is stored within the apparatus, so an external adversary cannot easily intercept the reflected signal.

[6] Strictly speaking, this latter prototype was a quantum lidar rather than a quantum radar, because its signal was transmitted at visible frequencies.

of both the target's range and direction (Maccone and Ren, 2020). Yet even with this new design, there is no way to track the target's speed or direction of movement, as existing radars can do via the Doppler effect. Thus, the military utility of quantum radar remains limited without significant progress in its design.

Quantum Communication

The primary near-term application of quantum communication technology is security against eavesdroppers, primarily through a method known as quantum key distribution (QKD). In QKD, an encryption key is transmitted between the two communicating parties in the form of quantum particles called photons. As a result of the quantum nature of these particles, any eavesdropper who intercepts them will, in principle, necessarily leave a signature on the data stream itself; if the protocol is implemented properly, then it is physically impossible to observe the photons without modifying them in a way. If the two communicating parties exchange an uncorrupted encryption key, then they can be assured that no one intercepted their key transmission and thus no one can decrypt the corresponding encrypted data. For this reason, quantum-secure communication is sometimes referred to as "unhackable."

However, this term oversimplifies the security of QKD. In practice, there are technical subtleties that can lead to vulnerabilities (Jain et al., 2014; Pang et al., 2020), and commercial implementations of QKD have been repeatedly demonstrated to have security vulnerabilities (Huang et al., 2016). Moreover, even if the communication channel's vulnerabilities to physical interception were secured, the signal would still need to be encoded and decoded at the endpoints by computers that are vulnerable to hacking, so securing the data stream against physical interception is not enough to ensure communication security (Schneier, 2008).

QKD has been demonstrated over three different physical communication channels: fiber optic cables, open air, and a satellite. Fiber optic cable is the most common medium, but the transmission endpoints must be fixed, and the photons can travel only a few hundred kilometers before degrading (Boaron et al., 2018). Transmission through free space allows for movable endpoints, but it requires direct line-of-sight transmission, and atmospheric interference currently limits the maximum distance to even shorter distances (Shmitt-Manderbach et al., 2007).[7] A much more dramatic demonstration of QKD came in 2017 and 2018, when the Chinese quantum satellite *Mozi* beamed streams of photons from space that encrypted a completely secure 75-minute teleconference between China and Austria (Liao et al., 2018). So far, *Mozi* is the only satellite to demonstrate QKD from space.

QKD over fiber optic cables is already in limited commercial use today. A Swiss company called ID Quantique has established QKD between Dutch telecom data centers, between Swiss banks, and between Swiss government election centers (Palmer, 2017; ID Quantique, 2017). The Chinese have built an extensive network connecting Beijing and Shanghai (Yiu, 2018), and, in 2020, Japanese researchers at Toshiba set a new record for data transmission via QKD by transmitting hundreds of gigabytes of human genomic data (Katsuda, 2020).

Quantum Computing

Quantum computing is the most widely known application of quantum technology but is also likely the furthest away from deployment (Monroe, 2019). A quantum computer could, in principle, perform certain computations vastly more quickly than is fundamentally possible with a standard (also referred to as a *classical*) computer—so much so that certain problems that are completely infeasible to solve on a standard computer could become feasible on a quantum computer. Despite this, there are surprisingly few known algorithms for specific applications. In fact, there are many computational problems for which quantum computers are *not* expected to deliver any significant improvement over classical computers (Aaronson, 2008).[8] Moreover, there are immense engineering challenges to building a useful

---

[7] The current distance records for QKD over fiber optic cable and through free space are 421 km and 144 km, respectively, but these were at the very edge of technological feasibility and had extremely low transmission rates.

[8] In particular, this includes the important class of computational problems known as NP-complete problems.

quantum computer, and only small-scale prototypes exist today (as will be discussed later).

The two most important quantum algorithms are Shor's algorithm and Grover's algorithm. Shor's algorithm (Shor, 1997) can be used to factor large numbers exponentially faster than any known classical algorithm.[9] Virtually all public-key encryption algorithms for protecting internet traffic rely on the computational difficulty of factoring and similar calculations.[10] The primary application of Shor's algorithm would therefore be the decryption of sensitive information transmitted over open channels, such as the internet, with obvious implications for online commerce and national security (discussed later). Shor's algorithm could also be used to destroy the security of most blockchain protocols (Fedorov, Kiktenko, and Lvovsky, 2018), including Bitcoin's, although there have been proposals for blockchain designs that are safe against quantum attacks (Kiktenko et al., 2018).

Grover's algorithm (Grover, 1996) improves the speed of brute-force search of a large database, as well as numerical optimization.[11] However, the speedup from Grover's algorithm is much less dramatic than from Shor's algorithm. Shor's algorithm gives an exponential speedup over the best-known classical algorithm, while Grover's algorithm gives only a square-root speedup.[12] Moreover, this speedup has been mathematically proven to be optimal for brute-force search; no possible quantum algorithm could

---

[9] More precisely, the speedup from Shor's algorithm scales as $\exp(n^{1/3})$, where n is the number of digits of the number to be factored. This speedup is technically not considered to be exponential but is enormous and often loosely described as exponential, even by experts.

[10] For example, three widely used public-key algorithms for encryption, key exchange, and digital signature are the Rivest-Shamir-Adleman (RSA), Diffie-Hellman, and elliptic-curve algorithms. The first algorithm depends on the computational difficulty of factoring, and the second two depend on the computational difficulty of the discrete logarithm problem. Shor's algorithm can be used to efficiently calculate both factors and discrete logarithms, so it destroys the effectiveness of all three algorithms.

[11] This is possible because any optimization problem can be reformulated as a search problem.

[12] More precisely, a classical computer requires $o(N)$ oracle queries to search an N-item database, while a quantum computer can use Grover's algorithm to search in $o(N^{1/2})$ queries.

deliver a better speedup. Nevertheless, even a modest improvement in database search and numerical optimization could have major applications in many fields that require high-performance computing, such as engineering and basic science (e.g., biochemistry and materials science). A very incomplete list of potential applications includes logistics, supply chain optimization, delivery routing, financial management, and the modeling of complex physical systems (such as nuclear or weather systems).

A third major algorithm known as the Harrow-Hassidim-Lloyd (HHL) algorithm was discovered more recently (Harrow, Hassidim, and Lloyd, 2009), and its significance is still an area of active research. The HHL algorithm efficiently performs certain linear-algebra calculations, which could potentially greatly improve machine-learning algorithms used in artificial intelligence. The discovery of the HHL algorithm has led to a recent boom in research in possible applications of quantum computers for machine learning (Havlíček et al., 2019; Schuld and Killoran, 2019), but many experts are still skeptical that quantum computers will offer a dramatic improvement in this area (Aaronson, 2015; Preskill, 2018).

A final, lesser-known potential application of quantum computers is for scientific simulation of advanced materials and biochemistry (Reiher et al., 2017; Wecker et al., 2015), including for drug discovery (Cao et al., 2018) and carbon capture (Preskill, 2018). Because quantum-mechanical effects explain the underlying physics of these materials, computers that use quantum mechanics are uniquely well suited to simulate them computationally. To give a sense of scale for the potential economic benefits, Reiher et al., 2017, gives a detailed theoretical proposal for using a modest-sized quantum computer to improve the efficiency of the Haber-Bosch process for industrial ammonia production, which currently consumes 2 percent of *world* energy production (primarily for making fertilizer). Even a fractional improvement in this process's efficiency could save billions of dollars per year.

Quantum Technologies That Span Categories

All of the near-term applications of quantum technology that are currently under development fall into one of the three categories of sensing, communication, and computing described earlier.[13] However, certain enabling technologies could spur progress across multiple categories. For example,

- A stable *quantum memory* is necessary for storing information in quantum computers over long periods but could also improve both the security of quantum communication technology (Liao et al., 2017) and the range of quantum radar (Barzanjeh et al., 2015).
- The *photonic qubits* used for long-distance quantum communication could also be used for quantum computing (although this type of qubit is not currently considered the most promising for quantum computers) (Ladd et al., 2010).

Technological progress will therefore likely not proceed completely independently among the three broad categories of quantum technology.

A longer-term application of quantum technology, known as *quantum networks* (or sometimes a *quantum internet*), could span all three of these categories. A quantum network of communication nodes could be very secure and could enable distributed quantum sensing and computing, among other applications that are difficult to predict today (Kimble, 2008).

---

[13] However, some sources treat long-range imaging technologies, such as ghost imaging and quantum radar, as a separate category from other quantum sensing technology.

CHAPTER THREE

Outlook over the Next Few Years

There is a simple (though technical) unifying measure, known as *quantum entanglement*, that quantifies both the performance and the engineering difficulty facing a given quantum technology. Roughly speaking, *entanglement* refers to multiple microscopic particles working together in a coordinated way that compounds their individual capabilities. The most-powerful potential applications of quantum technology all require sustained entanglement between many particles. However, such large-scale entanglement is *extremely* challenging to engineer because entanglement is very fragile, and preserving it requires keeping the particles extremely well isolated from their surroundings (and typically cooled down to a few thousandths of a degree above absolute zero).

Most quantum sensing applications require very little controlled entanglement, so they are considered the closest to commercial deployment. There is little information about potential military PNT applications in the open literature, but because the ghost imaging technology discussed in Chapter Two is quite mature (as demonstrated in Figure 2.1), it could possibly be fielded commercially within a few years. Quantum radar does require some entanglement and is likely much further away because certain basic science challenges still need to be overcome; one expert estimated that it might be fielded by 2030 (Seffers, 2015). The Chinese government claims to have already fielded a working quantum radar prototype, but experts are highly skeptical that this is true (Chen, 2016; Lin and Singer, 2018; Trimble, 2018).

Quantum sensing applications for navigation without GPS may also be close to practical usefulness. In 2015, an Air Force Scientific Advisory Board report concluded that quantum navigation sensors could be brought to Technology Readiness Level (TRL) 6 within the 2020–2025 time frame

(Air Force Scientific Advisory Board, 2015). Furthermore, in 2019, Lockheed Martin announced a prototype quantum magnetometer for navigation without GPS (Cameron, 2019). The U.S. start-up ColdQuanta is also working on what it calls quantum positioning systems and received U.S. Department of Defense (DoD) funding to develop some of the enabling technologies (ColdQuanta, undated; ColdQuanta, 2019). Although there are some commercial applications for inertial navigation (e.g., autonomous vehicles), the most important applications are military, so research and development (R&D) will likely be driven by the military both in the United States and in other countries (Air Force Scientific Advisory Board, 2015). The Five Eyes partnership (comprising Australia, Canada, New Zealand, the United Kingdom, and the United States) has announced a Quantum-Enabled PNT Strategic Challenge, whose goal is to use quantum sensors on a shipboard platform to demonstrate new capabilities for navigation in GPS-challenged environments by the year 2022 (Lawrence, 2019). In summary, quantum sensing technology could plausibly reach commercial or military maturity within the next few years.

Quantum communication is an intermediate case because some protocols use entanglement and some do not (although entanglement provides additional security). Quantum communication that is not based on entanglement has already been commercially deployed (in the form of QKD), but entanglement-based communication is still several years away. For example, the Chinese *Mozi* satellite has distributed pairs of entangled photons over a separation of 1,200 km (Yin et al., 2017), but the QKD protocol used to secure the intercontinental video teleconference described earlier did *not* use entanglement, so it was somewhat less secure than an entanglement-based protocol would have been.[1] Eventually, sophisticated networks of entanglement-based quantum communication could lead to the quantum internet described in the previous chapter. A full quantum internet would

---

[1] Quantum communication protocols that do not use entanglement typically require that the transmitted data be repeatedly decoded and re-encoded at intermediate points in the transmission chain (e.g., at signal boosters), and each of these points introduces a security vulnerability to undetected message interception. In the case of the Mozi satellite transmission, the satellite itself stored the encryption keys during its transit between the two ground stations in China and Austria, and it remained vulnerable to hacking during this transit.

require fundamental advances in both quantum communication and computing technology and is many years away, although small-scale implementations of quantum networks may come online in just a few years (Wehner, Elkouss, and Hanson, 2018). The first implementations will likely be used to extend the range of secure quantum communication beyond the limited range discussed earlier. In summary, the simplest forms of quantum communication have already been deployed, but the most-advanced and useful forms are still likely many years away.

Quantum computing is the most technologically challenging case because it requires a high degree of entanglement. As explained later in this chapter, all existing quantum computers are very, very far from being able to perform the applications described in Chapter Two.

The basic building block of a quantum computer is known as a *qubit*, which is related to the ordinary bits (0s and 1s) used by a classical computer. The power of a quantum computer can be roughly measured by the number of high-quality qubits that are entangled together inside of it. A quantum computer requires a minimum of about 50 high-quality qubits to perform some calculation that is too difficult for any existing supercomputer—a milestone sometimes referred to as *quantum supremacy*. In October 2019, Google asserted that it had achieved quantum supremacy with a 53-qubit quantum computer known as Sycamore. The Sycamore quantum computer took a few minutes to perform a calculation that Google claimed would take the world's fastest supercomputer more than 10,000 years (Arute et al., 2019).[2] However, the specific computational problem that Sycamore solved has no known practical applications; the computer was built primarily to provide a proof of principle that quantum computers can deliver a performance boost for certain problems, even if they are contrived (Boixo et al., 2018). Although the quantum supremacy milestone serves as an important

---

[2] IBM has questioned Google's claim that the fastest standard supercomputer would take 10,000 years to solve the problem and instead claims that it would take only 2.5 days (Cho, 2019). Without running the calculation on the supercomputer (which would be prohibitively expensive), it is difficult to assess who is correct. But even if IBM is correct, the addition of just a few more qubits to the Sycamore computer would cause it to far surpass any existing supercomputer.

proof of concept for quantum computers, it will not lead to any immediate commercial applications.[3]

By far the biggest technical challenge to scaling quantum computers to useful sizes is known as *quantum error correction*. All currently known designs for physical qubits have fairly high error rates, and the accumulation of errors quickly ruins the calculation.[4] This problem can, in principle, be solved by implementing quantum error correction, which is well understood theoretically but has just barely begun to be demonstrated experimentally (Ofek et al., 2016).

A major challenge of quantum error correction is that it requires an enormous hardware overhead, in the form of a huge number of qubits that simply correct errors rather than directly perform useful calculations. To give a sense of scale for this overhead, a paper published in 2021 estimates that a quantum computer could break commercial-grade 2048-bit RSA encryption with 14,238 hypothetical perfect qubits (Gidney and Ekerå, 2021). However, it would require 20 *million* imperfect qubits that could be realistically made with current technology. Because of these challenges, the authors of a 2019 National Academy of Sciences Consensus Study Report concluded, "It is highly unexpected that a quantum computer that can compromise RSA 2048 or comparable . . . public key cryptosystems will be built within the next decade" (National Academies of Sciences, Engineering, and Medicine, 2019). Applications for machine learning appear to face similarly formidable hardware requirements. The hardware requirements for applications for materials and biochemistry simulation are smaller (Wecker et al., 2014); although they will likely require some degree of error correction and

---

[3] Quantum computing experts have recently made two proposals for using these near-term quantum computers to generate certifiably random numbers (Ananthaswamy, 2019), but, as of this writing, neither proposal had yet been peer-reviewed, and it is not clear whether the proposals will prove useful.

[4] This problem is known as *decoherence*. Technically, decoherence refers to the destruction of the qubits' entanglement as a result of interaction with environmental noise. Google's Bristlecone computer can operate for only about 20 microseconds before its qubits decohere and need to be reset, which strongly limits the time for useful computation.

so are still several years away, they are likely to become practical significantly sooner than decryption will.[5]

Because the best-understood applications of quantum computers have
such formidable technical requirements, there are many proposals for algorithms that could run on computers that do not require error correction.
These *noisy intermediate-scale quantum* (NISQ) computers could be available within the next few years, and Preskill, 2018, gives an excellent nontechnical summary of their near-term prospects. There are two main takeaways:

1. The most-promising near-term algorithms are designed to run on
    hybrid quantum-classical computers instead of full quantum com-
    puters. There are such hybrid algorithms for combinatorial optimi-
    zation and materials simulation,[6] but they are not nearly as powerful
    as the algorithms described in Chapter Two.
2. It is very difficult to predict theoretically which of these near-term
    algorithms will be useful. The scientific community will need to
    build and test small quantum computers to find out.

Given the present low technological maturity of quantum computers,
some quantum industry trackers and start-up founders are concerned that
the current level of private start-up activity in quantum technology may
be excessive and premature and that unreasonably high expectations may
lead to a sudden crash in investment in a few years. This possibility has
been referred to as a *quantum winter*, by analogy with the past *AI winters*
in which research and investment in artificial intelligence technology has
suddenly dried up (Gibney, 2019). U.S. private-sector companies and the
popular media are generally more optimistic than academics are about the
near-term commercial potential of quantum computers, which may reflect

---

[5] For example, Reiher et al., 2017, estimates that useful chemistry simulations could be
performed with "only" 200,000 qubits. There are not yet any quantitative estimates for
the number of qubits required for machine-learning applications, but there are theoretical reasons to believe that they will require a much higher number of qubits.

[6] The combinatorial optimization algorithm is known as the Quantum Approximate
Optimization Algorithm, and the materials simulation algorithm is known as the Vari-
ational Quantum Eigensolver.

private companies' need to drum up investor enthusiasm (Preskill, 2018; Monroe, 2019).

In summary, some niche applications of quantum computers are being explored that may become useful within the next few years, but the most-important applications (such as breaking decryption) are likely at least ten years away.

CHAPTER FOUR

Major Players

Because quantum technology is a relatively small and new field, there have been few systematic international comparisons of investments, research quality, or overall technological leadership.

To get an idea of these factors, one simple metric to measure is a country's total public and private R&D investment; however, this does not correlate perfectly to technological leadership, especially in fields (such as quantum technology) that do not require huge capital expenditures. Table 4.1 is a snapshot of the estimated global R&D in quantum technologies in 2015. These numbers are significantly higher today. The United States, China, the EU, the United Kingdom, and Canada all have specific national initiatives to encourage quantum-related research (Crane et al., 2017; Figliola, 2019). The U.S. initiative, the National Quantum Initiative Act, is the most recent, having been signed into law in December 2018 (American Institute of Physics, 2019).

Until the early years of the 2010s, almost all investment in quantum R&D was government-funded. But there has been a large increase in private-sector investment that began around 2012 and accelerated around 2016 (Gibney, 2019). Figure 4.1 shows the total private-sector investment in quantum R&D by country from 2012 to 2018. The numbers in Table 4.1 and Figure 4.1 are not precisely comparable because Figure 4.1 covers a longer period, but the data demonstrate slightly different trends in total versus private investment. For example, the private-sector investments in Canada and the United States were comparable and were each larger than all of Europe's, Australia's private investment was comparable to Europe's, and China had relatively little private investment while Japan had almost none.

There is no clear world leader in quantum technology, because different countries excel in different subfields. The United States and China

TABLE 4.1

Total Investments in Quantum Technology Research, by Country, 2015

	Total Investment ($ millions)
World	1,665 (estimate)
Australia	83
Brazil	12
Canada	111
China	244
EU	610 (combined estimate)
Austria	39
Denmark	24
Finland	13
France	58
Germany	133
Great Britain	117
Italy	40
Netherlands	30
Poland	13
Spain	28
Sweden	17
Japan	70
Russia	33
Singapore	49
South Korea	14
Switzerland	74
United States	400

SOURCE: Palmer, 2017.

NOTE: The numbers in this table indicate raw spending unadjusted for GDP.

FIGURE 4.1

Private-Sector Investments in Quantum Technology Research and Development, 2012–2018 ($U.S. millions)

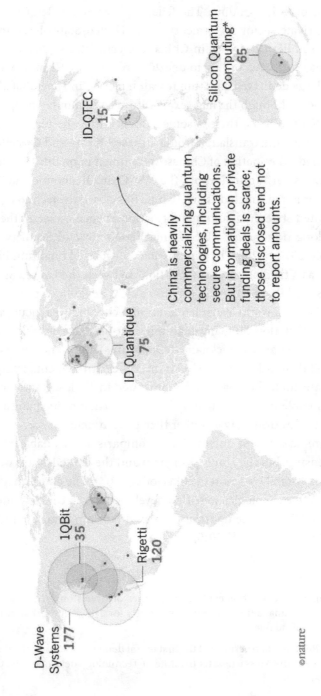

D-Wave Systems **177**

1QBit **35**

Rigetti **120**

ID Quantique **75**

ID-QTEC **15**

Silicon Quantum Computing* **65**

China is heavily commercializing quantum technologies, including secure communications. But information on private funding deals is scarce; those disclosed tend not to report amounts.

©nature

SOURCE: Adapted by permission from Springer Nature: Elizabeth Gibney, "Quantum Gold Rush: The Private Funding Pouring into Quantum Start-Ups," *Nature*, Vol. 574, © 2019, p. 23.

NOTE: * The Silicon Quantum Computing number includes an unspecified contribution from the Australian government alongside private investors. ID-QTEC = China Quantum Technologies.

dominate in overall spending: The federal government R&D investment (excluding private-sector investment) in the United States is estimated to be $200–250 million per year; in China, the estimate is $244 million per year (Figliola, 2018). According to one 2017 report, the Chinese were constructing a large dedicated quantum research facility in Hefei that allegedly cost $10 billion, although the huge expenditure could not be independently confirmed (Chen, 2017); there is some evidence that the report may have originated from a mistranslation from Chinese.[1] Kania and Costello, 2018, gives a detailed examination of Chinese investment in quantum technology as of 2018. In its 2017 annual report, the U.S.-China Economic and Security Review Commission concluded that "China has closed the technological gap with the United States in quantum information science—a sector the United States has long dominated" (U.S.-China Economic and Security Review Commission, 2017, p. 25). The commission categorized quantum information science as a field of "close competition" between the United States and China and stated that neither country has a clear advantage (p. 532).

Canada is also a world leader in quantum technology, particularly at the University of Waterloo. So many quantum start-ups have developed around Waterloo that the area is nicknamed Quantum Valley, and an associated venture capital fund known as Quantum Valley Investments is dedicated solely to quantum technology. As demonstrated in Table 4.1 and Figure 4.1, Canada is notable in that a relatively high proportion of its quantum R&D investment comes from private rather than government sources.

There are few recent international comparisons available regarding quantum sensing, but the United Kingdom and the United States were considered international leaders as of 2014 (Jones, 2014). The ghost imaging and quantum radar technologies were first developed in the United States,[2] and researchers at Waterloo demonstrated the first tabletop quantum radar prototype in 2019 (Luong et al., 2019).

---

[1] Elsa B. Kania presented this evidence in a series of tweets on Twitter, starting on July 12, 2018 (see Kania, 2018). I searched extensively for a more authoritative reference but was unable to find one.

[2] The Army Research Lab performed the first useful demonstrations of ghost imaging, and researchers at the Massachusetts Institute of Technology developed the theory of quantum radar.

In quantum communication, China is the clear world leader (Yiu, 2018). As discussed in Chapter Two, China is the only country in the world to have fielded a satellite capable of quantum communication with the ground, and the Chinese have built an extensive fiber optic network for QKD between Beijing and Shanghai. Much of China's success is associated with a single scientist, Jian-Wei Pan, who has been involved with nearly all of China's technological breakthroughs in quantum communication and photonics (Giles, 2018). Pan leads the new quantum research facility in Hefei mentioned before.

In quantum computing, the United States is the clear world leader. The largest quantum computers (with 50 or more high-quality qubits) have been built by Google (Kelly, 2018), IBM (IBM, 2019), and a start-up named IonQ (IonQ, 2018). Other major U.S. commercial players include Honeywell; Microsoft; and a variety of start-ups, such as Rigetti, ColdQuanta, and PsiQuantum. Another notable quantum computing group is the Canadian company D-Wave systems from Waterloo. Google, Lockheed Martin, and the National Aeronautics and Space Administration (NASA) have purchased D-Wave machines with more than 2,000 qubits; however, many academic researchers in quantum computing are skeptical of the D-Wave computer's performance because its qubits are of much lower quality than the qubits in the other computers (Denchev et al., 2016).[3] Although the Chinese firm Alibaba is investing in quantum computing, it has not announced computers with qubit counts (or any other quality metrics) nearly as high as U.S. companies' computers. Unlike in the United States, nearly all Chinese investment in quantum computing appears to come from the government rather than from private companies (Crane et al., 2017).

---

[3] The other companies mentioned in this paragraph are taking an approach known as *gate-based quantum computing*, while D-Wave's quantum computer uses a fundamentally different paradigm known as *adiabatic quantum computing*. This makes the companies' quantum computers somewhat difficult to compare.

CHAPTER FIVE

National Security Implications

Every subfield of quantum technology potentially has major implications for national security (as well as the private sector), although the time frames for those implications differ (Giles, 2019).

Most of the national security–related applications of quantum sensing are described in Chapter Two. The primary applications are for ISR and PNT. Quantum radar could, in principle, increase maximum radar range by 41 percent and could be particularly effective against stealth aircraft. However, this capability is likely many years away and may never come to fruition; the Defense Science Board has concluded, "Quantum radar will not provide upgraded capability to DoD" (Defense Science Board, 2019, p. 2). In the nearer term, ghost imaging technology could improve air-based ISR through cloud cover and smoke. Sensitive atomic clocks could improve GPS-based positioning, and magnetometers and gravitometers could allow for navigation in GPS-denied environments via the Earth's fields. Accelerometers could improve inertial navigation systems, including those in guided missiles. The Air Force has concluded that, with sustained development, these technologies could be mature in the next few years (Air Force Scientific Advisory Board, 2015).

In the near term, quantum communication technology could use QKD to protect sensitive encrypted communications (military, governmental, or commercial) against hostile interception. This is the application that the Chinese government is focusing on. On the other hand, U.S. researchers have not focused on QKD. Moreover, the Air Force has concluded that QKD is unlikely to provide a significant advantage over current capabilities (Air Force Scientific Advisory Board, 2015), and the Defense Science Board (2019, p. 4) found that "QKD has not been implemented with sufficient capability or security to be deployed for DoD mission use." The United Kingdom's Government Communication Headquarters has also recommended against the government or military adoption of QKD (National Cyber Security Centre, 2018). The U.S. National Security Agency concurs with this assessment and has publicly stated that it "does not support the usage of QKD . . . to protect communications in National Security Systems" (U.S. National Security Agency Central Security Service, undated). This is because QKD introduces new complexity (and therefore potential vulnerabilities) into the communication chain. Quantum-secured information will still remain vulnerable to the exploitation of other weak links (e.g., software vulnerabilities at the endpoints), just as they are today (Schneier, 2008). Moreover, as discussed earlier, commercial QKD devices have repeatedly been demonstrated to contain security vulnerabilities (Jain et al., 2014; Huang et al., 2016; Pang et al., 2020). The fact that China, Europe, and Japan are moving quickly to deploy QKD while the U.S. and United Kingdom defense communities are publicly discouraging its deployment indicates these groups of nations' two divergent schools of thought about QKD's usefulness.

Theoretically, quantum computing could eventually have the most-severe impact on national security. A large-scale quantum computer capable of deploying Shor's algorithm on current commercial-grade encryption (such as 2048-bit RSA) would have a devastating impact on virtually all internet security. Without reliable encryption, private communication online would become impossible, which would make online commerce and other financial transactions impossible (without a means to securely transmit payment information) and would render email useless for sensitive communications.[1]

But quantum computers capable of running Shor's algorithm are almost certainly more than ten years away (National Academies of Sciences, Engi-

---

[1] It is worth mentioning that quantum computers are primarily effective against *public-key encryption*, which is used when the two communication parties have no way of securely exchanging an encryption key offline. When the parties can physically and securely exchange a key, they can instead use *symmetric-key encryption* algorithms. The strongest of these algorithms are believed to be vulnerable only to Grover's algorithm, which is much less powerful than Shor's algorithm. When the parties can securely exchange an encryption key, the vulnerability to quantum attacks is serious but not devastating.

neering, and Medicine, 2019). Moreover, the U.S. government is already developing a response to the threat posed by quantum computers. Specifically, the National Institute of Standards and Technology is developing new encryption algorithms that are believed to be secure against attacks from future quantum computers (Alagic et al., 2019). A 2020 RAND report discusses the policy implications of the transition to these new forms of cryptography (Vermeer and Peet, 2020). Although the transition to new algorithms will be disruptive, the development of large-scale quantum computers is unlikely to permanently destroy the feasibility of public-key cryptography.

In the near term, the smaller-scale NISQ computers discussed in Chapter Three do not have any obvious direct effects on national security. Tasks that are relevant to the military and the intelligence community (such as target optimization and machine learning) are likely too difficult for NISQ computers (Preskill, 2018). Any impact on national security will most likely be only indirect and will result via general economic benefits from improved (publicly available) scientific and biomedical knowledge.

CHAPTER SIX

Conclusion

The near-term impacts of quantum computers are sometimes exaggerated; existing quantum computers will need to be scaled up by a factor of as much as 1 million before they become relevant for cryptography. Near-term quantum computers do not yet have any guaranteed applications, but simulation for scientific research looks promising and could have a major economic impact. Quantum sensors are much closer to commercial deployment and could provide significant improvements in long-distance imaging, timing, and sensing of electric, magnetic, and gravitational fields, with both commercial and military applications. QKD could, in principle, improve the security of communications against eavesdropping, but many U.S. and United Kingdom experts are skeptical that it will prove useful in practice. Further into the future, more-sophisticated forms of quantum communication technology could be used to network together quantum computers and sensors, but these networks are still at a very early stage of development.

The United States, closely followed by Canada, the United Kingdom, and the EU, are the world leaders in quantum sensing and computing. China has chosen to focus its research efforts on quantum communication instead and is widely considered to be the world leader in that subfield, because it is the only country to have launched a QKD satellite and has laid down the world's largest network of fiber optic cables for QKD. Neither the United States nor China has a clear lead in quantum technology overall; the U.S.-China Economic and Security Review Commission (2017, p. 532) has characterized quantum information science as an area of "close competition" between the two countries.

Abbreviations

DoD	U.S. Department of Defense
EU	European Union
GPS	Global Positioning System
HHL	Harrow-Hassidim-Lloyd [quantum algorithm]
INS	inertial navigation system
ISR	intelligence, surveillance, and reconnaissance
NISQ	noisy intermediate-scale quantum
PNT	positioning, navigation, and timing
QIS	quantum information science
QKD	quantum key distribution
R&D	research and development
RSA	Rivest-Shamir-Adleman [encryption system]
SNR	signal-to-noise ratio

References

Aaronson, Scott, "The Limits of Quantum Computers," *Scientific American*, March 1, 2008.

———, "Read the Fine Print," *Nature Physics*, Vol. 11, April 2015, pp. 291–293.

Air Force Scientific Advisory Board, *Utility of Quantum Systems for the Air Force*, Washington, D.C., 2015.

Alagic, Gorjan, Jacob Alperin-Sheriff, Daniel Apon, David Cooper, Quynh Dang, Carl Miller, Dustin Moody, Rene Peralta, Ray Perlner, Angela Robinson, Daniel Smith-Tone, and Yi-Kai Liu, *NISTIR 8240: Status Report on the First Round of the NIST Post-Quantum Cryptography Standardization Process*, Gaithersburg, Md.: National Institute of Standards and Technology, January 2019.

American Institute of Physics, "National Quantum Initiative Act, H.R.6227/S.3143," webpage, 2019. As of July 30, 2020: https://www.aip.org/fyi/federal-science-bill-tracker/115th/national-quantum-initiative-act

Ananthaswamy, Anil, "How to Turn a Quantum Computer into the Ultimate Randomness Generator," *Quanta Magazine*, June 19, 2019.

Army Research Laboratory, "Army Scientists' 19 Patents Lead to Quantum Imaging Advances," December 19, 2013.

Arute, Frank, Kunal Arya, Ryan Babbush, Dave Bacon, Joseph C. Bardin, Rami Barends, Rupak Biswas, Sergio Boixo, et al., "Quantum Supremacy Using a Programmable Superconducting Processor," *Nature*, No. 574, 2019, pp. 505–510.

Aspden, Reuben S., Nathan R. Gemmell, Peter A. Morris, Daniel S. Tasca, Lena Mertens, Michael G. Tanner, Robert A. Kirkwood, Alessandro Ruggeri, Alberto Tosi, Robert W. Boyd, Gerald S. Buller, Robert H. Hadfield, and Miles J. Padgett, "Photon-Sparse Microscopy: Visible Light Imaging Using Infrared Illumination," *Optica*, Vol. 2, No. 12, 2015, pp. 1049–1052.

Ball, Philip, "Quantum Mechanics Could Improve Radar," *Physics*, Vol. 8, 2015, p. 18.

Barzanjeh, Shabir, Saikat Guha, Christian Weedbrook, David Vitali, Jeffrey H. Shapiro, and Stefano Pirandola, "Microwave Quantum Illumination," *Physical Review Letters*, Vol. 114, No. 8, 2015, pp. 1–5.

Boaron, Alberto, Gianluca Boso, Davide Rusca, Cédric Vulliez, Claire Autebert, Misael Caloz, Matthieu Perrenoud, Gaëtan Gras, Félix Bussières, Ming-Jun Li, Daniel Nolan, Anthony Martin, and Hugo Zbinden, "Secure Quantum Key Distribution over 421 km of Optical Fiber," *Physical Review Letters*, Vol. 121, No. 19, 2018.

Boixo, Sergio, Sergei V. Isakov, Vadim N. Smelyanskiy, Ryan Babbush, Nan Ding, Zhang Jiang, Michael J. Bremner, John M. Martinis, and Hartmut Neven, "Characterizing Quantum Supremacy in Near-Term Devices," *Nature Physics*, Vol. 14, No. 6, 2018, pp. 595–600.

Bowler, Tim, "How Quantum Sensing Is Changing the Way We See the World," BBC News, March 8, 2019.

Cameron, Alan, "Quantum Magnetometer Senses Its Place," GPS World, May 8, 2019.

Cao, Yudong, Jhonathan Romero Fontalvo, and Alán Aspuru-Guzik, "Potential of Quantum Computing for Drug Discovery," *IBM Journal of Research and Development*, Vol. 62, No. 6, November/December 2018.

Chen, Stephen, "The End of Stealth? New Chinese Radar Capable of Detecting 'Invisible' Targets 100km Away," *South China Morning Post*, September 21, 2016.

———, "China Building World's Biggest Quantum Research Facility," *South China Morning Post*, September 11, 2017.

Cho, A., "IBM Casts Doubt on Google's Claims of Quantum Supremacy," *Science*, October 23, 2019.

ColdQuanta, homepage, undated. As of July 30, 2021:
https://www.coldquanta.com/

———, "ColdQuanta Awarded $2.8M from the U.S. Government to Advance Its Quantum Core Technology," press release, BusinessWire, October 29, 2019.

Crane, Keith W., Lance G. Joneckis, Hannah Acheson-Field, Iain D. Boyd, Benjamin A. Corbin, Xueying Han, and Robert N. Rozansky, *Assessment of the Future Economic Impact of Quantum Information Science*, Washington, D.C.: Institute for Defense Analyses, Science and Technology Policy Institute, August 2017.

Defense Science Board, *Applications of Quantum Technologies: Executive Summary*, Washington, D.C.: Office of the Under Secretary of Defense for Research and Engineering, October 2019.

Degen, C. L., F. Reinhard, and P. Cappellaro, "Quantum Sensing," *Reviews of Modern Physics*, Vol. 89, No. 3, 2017.

Denchev, Vasil S., Sergio Boixo, Sergei V. Isakov, Nan Ding, Ryan Babbush, Vadim Smelyanskiy, John Martinis, and Hartmut Neven, "What Is the Computational Value of Finite-Range Tunneling?" *Physical Review X*, Vol. 6, No. 3, 2016.

Fedorov, Aleksey K., Evgeniy O. Kiktenko, and Alexander I. Lvovsky, "Quantum Computers Put Blockchain Security at Risk," *Nature*, Vol. 563, 2018, pp. 465–467.

Figliola, Patricia Moloney, *Federal Quantum Information Science: An Overview*, Washington, D.C.: Congressional Research Service, July 2, 2018.

———, *Quantum Information Science: Applications, Global Research and Development, and Policy Considerations*, Washington, D.C.: Congressional Research Service, R45409, 2019.

Genovese, Marco, "Real Applications of Quantum Imaging," *Journal of Optics*, Vol. 18, No. 7, 2016.

Gibney, Elizabeth, "Quantum Gold Rush: The Private Funding Pouring into Quantum Start-Ups," *Nature*, Vol. 574, 2019, pp. 22–24.

Gidney, Craig, and Martin Ekerå, "How to Factor 2048 Bit RSA Integers in 8 Hours Using 20 Million Noisy Qubits," *Quantum*, Vol. 5, April 2021, pp. 433–464.

Giles, Martin, "The Man Turning China into a Quantum Superpower," *MIT Technology Review*, December 19, 2018.

———, "The US and China Are in a Quantum Arms Race That Will Transform Warfare," *MIT Technology Review*, January 3, 2019.

Grover, Lov K., "A Fast Quantum Mechanical Algorithm for Database Search," *28th Annual ACM Symposium on the Theory of Computing*, 1996.

Guha, Saikat, and Baris I. Erkmen, "Gaussian-State Quantum-Illumination Receivers for Target Detection," *Physical Review A*, Vol. 80, No. 5, 2009.

Hardy, Nicholas D., and Jeffrey H. Shapiro, "Computational Ghost Imaging Versus Imaging Laser Radar for Three-Dimensional Imaging" *Physical Review A*, Vol. 87, No. 2, 2013.

Harrow, Aram W., Avinatan Hassidim, and Seth Lloyd, "Quantum Algorithm for Linear Systems of Equations," *Physical Review Letters*, Vol. 103, No. 15, 2009.

Havlíček, Vojtěch, Antonio D. Córcoles, Kristan Temme, Aram W. Harrow, Abhinav Kandala, Jerry M. Chow, and Jay M. Gambetta, "Supervised Learning with Quantum-Enhanced Feature Spaces," *Nature*, Vol. 567, No. 7747, 2019, pp. 209–212.

Huang, Anqi, Shihan Sajeed, Poompong Chaiwongkhot, Mathilde Soucarros, Matthieu Legré, and Vadim Makarov, "Testing Random-Detector-Efficiency Countermeasure in a Commercial System Reveals a Breakable Unrealistic Assumption," *IEEE Journal of Quantum Electronics*, Vol. 52, No. 11, November 2016.

IBM, "IBM Unveils World's First Integrated Quantum Computing System for Commercial Use," press release, January 8, 2019.

ID Quantique, "IDQ Celebrates 10-Year Anniversary of the World's First Real-Life Quantum Cryptography Installation," press release, November 23, 2017.

IonQ, "IonQ Harnesses Single-Atom Qubits to Build the World's Most Powerful Quantum Computer," press release, December 11, 2018.

Jain, Nitin, Elena Anisimova, Imran Khan, Vadim Makarov, Christoph Marquardt, and Gerd Leuchs, "Trojan-Horse Attacks Threaten the Security of Practical Quantum Cryptography," *New Journal of Physics*, Vol. 16, No. 12, 2014.

Jones, Sam, "MoD's 'Quantum Compass' Offers Potential to Replace GPS," *Financial Times*, May 14, 2014.

Kania, Elsa B. [EBKania], "Just a quick public service announcement, the Chinese government has *not* announced that it is spending $150 billion on AI, and China's National Laboratory for Quantum Information Science has *not* received $10 billion for funding...," Twitter post, July 12, 2018. As of August 3, 2020:
https://twitter.com/EBKania/status/1017604097694949377

Kania, Elsa B., and John Costello, *Quantum Hegemony? China's Ambitions and the Challenge to U.S. Innovation Leadership*, Washington, D.C.: Center for a New American Security, September 12, 2018.

Katsuda, Toshihiko, "Japan Test Makes Leap in Quantum Cryptography Development," *Asahi Shimbun*, January 14, 2020.

Kelly, J., "A Preview of Bristlecone, Google's New Quantum Processor," *Google AI Blog*, March 5, 2018.

Kiktenko, E. O., N. O. Pozhar, M. N. Anufriev, A. S. Trushechkin, R. R. Yunusov, Y. V. Kurochkin, A. I. Lvovsky, and A. K. Fedorov, "Quantum-Secured Blockchain," *Quantum Science and Technology*, Vol. 3, No. 3, 2018.

Kimble, H. J., "The Quantum Internet," *Nature*, Vol. 453, 2008, pp. 1023–1030.

Kramer, David, "DARPA Looks Beyond GPS for Positioning, Navigating, and Timing," *Physics Today*, Vol. 67, No. 19, 2014, pp. 23–26.

Ladd, T. D., F. Jelezko, R. Laflamme, Y. Nakamura, C. Monroe, and J. L. O'Brien, "Quantum Computers," *Nature*, Vol. 464, 2010, pp. 45–53.

Lawrence, Timothy, "Quantum Information Science: The Way Ahead," presentation slides from the 1st International Quantum Information Sciences Workshop at SUNY Polytechnic Institute, Utica, New York, July 9–10, 2019. As of July 30, 2020: https://www.suny.edu/media/suny/content-assets/images/research/events/AFRL-Quantum-Information-Science-The-Way-Ahead-Lawrence.pdf

Liao, Sheng-Kai, Wen-Qi Cai, Wei-Yue Liu, Liang Zhang, Yang Li, Ji-Gang Ren, Juan Yin, et al., "Satellite-to-Ground Quantum Key Distribution," *Nature*, Vol. 549, 2017, pp. 43–47.

Liao, Sheng-Kai, Wen-Qi Cai, Johannes Handsteiner, Bo Liu, Juan Yin, Liang Zhang, Dominik Rauch, et al., "Satellite-Relayed Intercontinental Quantum Network," *Physical Review Letters*, Vol. 120, No. 3, 2018.

Lin, Jeffrey, and P. W. Singer, "China's Latest Quantum Radar Could Help Detect Stealth Planes, Missiles," *Popular Science*, July 11, 2018.

Lloyd, Seth, "Enhanced Sensitivity of Photodetection via Quantum Illumination," *Science*, Vol. 321, No. 5895, 2008, pp. 1463–1465.

Luong, David, Anthony Damini, Bhashyam Balaji, C. W. Sandbo Chang, A. M. Vadiraj, and Christopher Wilson, "A Quantum-Enhanced Radar Prototype," 2019 IEEE Radar Conference, April 22–26, 2019.

Maccone, Lorenzo, and Changliang Ren, "Quantum Radar," *Physical Review Letters*, Vol. 124, No. 20, May 2020.

Marks, Paul, "Quantum Positioning System Steps in When GPS Fails," *New Scientist*, May 14, 2014.

Meyers, Ron, Keith S. Deacon, and Yanhua Shih, "Ghost-Imaging Experiment by Measuring Reflected Photons," *Physical Review A*, Vol. 77, No. 4, 2008.

Meyers, Ronald E., Keith S. Deacon, Arnold D. Tunick, and Yanhua Shih, "Virtual Ghost Imaging Through Turbulence and Obscurants Using Bessel Beam Illumination," *Applied Physics Letters*, Vol. 100, No. 6, 2012.

Monroe, Christopher, "Quantum Computing Is a Marathon Not a Sprint," *VentureBeat*, April 21, 2019.

Morris, Peter A., Reuben S. Aspden, Jessica E. C. Bell, Robert W. Boyd, and Miles J. Padgett, "Imaging with a Small Number of Photons," *Nature Communications*, Vol. 6, 2015, p. 5913.

National Academies of Sciences, Engineering, and Medicine, *Quantum Computing: Progress and Prospects*, Washington, D.C.: National Academies Press, 2019.

National Cyber Security Centre, *Quantum Key Distribution*, London, November 30, 2018.

Ofek, Nissim, Andrei Petrenko, Reinier Heeres, Philip Reinhold, Zaki Leghtas, Brian Vlastakis, Yehan Liu, Luigi Frunzio, S. M. Girvin, L. Jiang, Mazyar Mirrahimi, M. H. Devoret, and R. J. Schoelkopf, "Extending the Lifetime of a Quantum Bit with Error Correction in Superconducting Circuits," *Nature*, Vol. 536, 2016, pp. 441–445.

Palmer, Jason, "Here, There and Everywhere: Quantum Technology Is Beginning to Come into Its Own," *The Economist*, 2017.

Pang, Xiao-Ling, Ai-Lin Yang, Chao-Ni Zhang, Dou Jianpeng, Hang Li, Jun Gao, and Xian-Min Jin, "Hacking Quantum Key Distribution via Injection Locking," *Physical Review Applied*, Vol. 13, No. 3, 2020.

Pirandola, S., B. R. Bardhan, T. Gehring, C. Weedbrook, and S. Lloyd, "Advances in Photonic Quantum Sensing," *Nature Photonics*, Vol. 12, No. 12, 2018, pp. 724–733.

Preskill, John, "Quantum Computing in the NISQ Era and Beyond," *Quantum*, Vol. 2, 2018.

Reiher, Markus, Nathan Wiebe, Krysta Marie Svore, Dave Wrecker, and Matthias Troyer, "Elucidating Reaction Mechanisms on Quantum Computers," *Proceedings of the National Academy of Sciences*, Vol. 114, No. 29, 2017.

Schneier, Bruce, "Quantum Cryptography: As Awesome as It Is Pointless," *Wired*, October 15, 2008.

Schuld, M., and N. Killoran, "Quantum Machine Learning in Feature Hilbert Spaces," *Physical Review Letters*, Vol. 122, No. 4, 2019.

Seffers, George I., "Quantum Radar Could Render Stealth Aircraft Obsolete," *Signal*, July 1, 2015.

Shapiro, Jeffrey H., "Computational Ghost Imaging," *Physical Review A*, Vol. 78, No. 6, 2008.

———, "Defeating Passive Eavesdropping with Quantum Illumination," *Physical Review A*, Vol. 80, 2009.

Shmitt-Manderbach, Tobias, HenningWeier, Martin Fürst, Rupert Ursin, Felix Tiefenbacher, Thomas Scheidl, Josep Perdigues, Zoran Sodnik, Christian Kurtsiefer, John G. Rarity, Anton Zeilinger, and Harald Weinfurter, "Experimental Demonstration of Free-Space Decoy-State Quantum Key Distribution over 144 km," *Physical Review Letters*, Vol. 98, 2007.

Shor, Peter W., "Polynomial-Time Algorithms for Prime Factorization and Discrete Logarithms on a Quantum Computer," *SIAM Journal on Computing*, Vol. 26, No. 5, 1997.

Tan, Si-Hui, Baris I. Erkmen, Vittorio Giovannetti, Saikat Guha, Seth Lloyd, Lorenzo Maccone, Stefano Pirandola, and Jeffrey H. Shapiro, "Quantum Illumination with Gaussian States," *Physical Review Letters*, Vol. 101, No. 25, 2008.

Trimble, Steve, "China Shows Off First Quantum Radar Prototype," Aerospace Daily & Defense Report, November 5, 2018.

Tucker, Patrick, "Four DARPA Projects That Could Be Bigger Than the Internet," *Defense One*, May 20, 2014.

U.S.-China Economic and Security Review Commission, *2017 Report to Congress of the U.S.-China Economic and Security Review Commission*, Washington, D.C.: U.S. Government Publishing Office, 2017.

U.S. National Security Agency Central Security Service, "Quantum Key Distribution (QKD) and Quantum Cryptography (QC)," webpage, undated. As of August 3, 2021:
https://www.nsa.gov/What-We-Do/Cybersecurity/
Quantum-Key-Distribution-QKD-and-Quantum-Cryptography-QC/

Vermeer, Michael J. D., and Evan D. Peet, *Securing Communications in the Quantum Computing Age: Managing the Risks to Encryption*, Santa Monica, Calif.: RAND Corporation, RR-3102-RC, 2020. As of July 30, 2020:
https://www.rand.org/pubs/research_reports/RR3102.html

Wecker, Dave, Bela Bauer, Bryan K. Clark, Matthew B. Hastings, and Matthias Troyer, "Gate-Count Estimates for Performing Quantum Chemistry on Small Quantum Computers," *Physical Review A*, Vol. 90, No. 2, 2014.

Wecker, Dave, Matthew B. Hastings, Nathan Wiebe, Bryan K. Clark, Chetan Nayak, and Matthias Troyer, "Solving Strongly Correlated Electron Models on a Quantum Computer," *Physical Review A*, Vol. 92, No. 6, 2015.

Wehner, Stephanie, David Elkouss, and Ronald Hanson, "Quantum Internet: A Vision for the Road Ahead," *Science*, Vol. 362, No. 6412, 2018.

Yin, Juan, Yuan Cao, Yu-Huai Li, Sheng-Kai Liao, Liang Zhang, Ji-Gang Ren, Wen-Qi Cai, et al., "Satellite-Based Entanglement Distribution over 1200 Kilometers," *Science*, Vol. 356, No. 6343, 2017, pp. 1140–1144.

Yiu, Y., "Is China the Leader in Quantum Communications?" *Inside Science*, January 19, 2018.

Zhang, Zheshen, Sara Mouradian, Franco N.C. Wong, and Jeffrey H. Shapiro, "Entanglement-Enhanced Sensing in a Lossy and Noisy Environment," *Physical Review Letters*, Vol. 114, No. 11, 2015.

Zhao, Chengqiang, Wenlin Gong, Mingliang Chen, Enrong Li, Hui Wang, Wendong Xu, and Shensheng Han, "Ghost Imaging Lidar via Sparsity Constraints," *Applied Physics Letters*, Vol. 101, No. 14, 2012, pp. 141123-1–141123-3.